BLOG WILD!

BLOG WILD!

A GUIDE FOR SMALL BUSINESS BLOGGING

Andy Wibbels

PORTFOLIO

PORTFOLIO
Published by the Penguin Group
Penguin Group (USA) Inc., 375 Hudson Street, New York, New York 10014, U.S.A. · Penguin Group (Canada), 90 Eglinton Avenue East, Suite 700, Toronto, Ontario, Canada M4P 2Y3 (a division of Pearson Penguin Canada Inc.) · Penguin Books Ltd, 80 Strand, London WC2R 0RL, England · Penguin Ireland, 25 St. Stephen's Green, Dublin 2, Ireland (a division of Penguin Books Ltd) · Penguin Books Australia Ltd, 250 Camberwell Road, Camberwell, Victoria 3124, Australia (a division of Pearson Australia Group Pty Ltd) · Penguin Books India Pvt Ltd, 11 Community Centre, Panchsheel Park, New Delhi –110 017, India · Penguin Group (NZ), Cnr Airborne and Rosedale Roads, Albany, Auckland 1310, New Zealand (a division of Pearson New Zealand Ltd) · Penguin Books (South Africa) (Pty) Ltd, 24 Sturdee Avenue, Rosebank, Johannesburg 2196, South Africa

Penguin Books Ltd, Registered Offices:
80 Strand, London WC2R 0RL, England

First published in 2006 by Portfolio,
a member of Penguin Group (USA) Inc.

1 3 5 7 9 10 8 6 4 2

Publisher's Note: This publication is designed to provide accurate and authoritative information in regard to the subject matter covered. It is sold with the understanding that the publisher is not engaged in rendering legal, accounting or other professional services. If you require legal advice or other expert assistance, you should seek the services of a competent professional.

While the author has made every effort to provide accurate telephone numbers and Internet addresses at the time of publication, neither the publisher nor the author assumes any responsibility for errors, or for changes that occur after publication. Further, the publisher does not have any control over and does not assume any responsibility for author or third-party Web sites or their content.

ISBN 1-59184-117-8
CIP data available

Printed in the United States of America
Set in Serifa • Designed by Jaime Putorti

To my parents,
who bought me that Commodore 64.

Special Thanks

Special thanks to Megan Casey (for getting things started), Joe Veltre (for guiding the way), Adrian Zackheim (for giving the green light), and Adrienne Schultz (for keeping the ball rolling), as well as all the great folks at Portfolio.

Huge thanks to all of my clients and colleagues who have helped me spread the gospel of blogging, including Mitch Meyerson, Michael Port, Bea Fields, Lisa Wilder, Lisa Micklin, Andrea J. Lee, Tina Forsyth, Susan Austin, Scott Stratten, Maryam Webster, Pat Gundry, Barb Elgin, Des Walsh, Bob Brannan, Rob Schultz, Eric Schneider, Darren Rowse, Duncan Riley, Peter Flaschner, Tony Marino, Wayne Kelly, Bob Sommers, Chris Pirillo, Steve Rubel, and the eagle eyes (and eternal patience) of Tara Kachaturoff.

And, of course, Mom, Dad, Heather, Brooks, and Ron.

Contents

Part Four:
Advanced Techniques

Part Five:
Promoting Your Blog

Introduction

Five thousand hits in one day. I was stunned.

In the midst of the 2004 U.S. Presidential Election, I started thinking to myself, "John Kerry looks really tired. You'd think that his army of stylists would be able to make him look a little warmer, a little less worn-out." Mind you, this was before the Botox injections. So I went to Google and searched for the phrase "John Kerry looks like" to see what other people thought the Democratic candidate for president looked like. A slew of opinions came back—mostly from bloggers and message boards. But one of them stuck out to me: "He looked like the Herman Munster character from the 1960's sitcom *The Munsters*."

To confirm my suspicions, I grabbed images of both Kerry and Munster and cooked up a little animation of John Kerry morphing into Herman Munster and posted it to my blog.[1]

A couple of friends and fellow bloggers saw it and laughed.

Over the next three weeks, more and more people started linking to my little cartoon. Gradually, more and more people found the link, and more and more people started visiting my blog. Then some high-traffic political blogs caught on, and my Web site was blitzed with traffic for five days solid. I was getting thousands of hits from people searching Google for "John Kerry Herman Munster." I received comments from all over the country, with people from all around the world laughing at the cartoon (or ranting about it).

That's when I started to investigate how entrepreneurs and businesses could leverage blogs to market their products and reach an audience hungry for information and entertainment. I started the Easy Bake Weblogs seminar (www.easybakeweblogs.com) and began sharing what I knew with individuals and companies all over the world. This book is based on that seminar.

Blogwild is for businesses and entrepreneurs who

---

[1]To see what I'm talking about, go to http://www.goblogwild.com/johnkerry/

have heard about blogs, know that there's something special going on, but aren't quite sure what the big deal is. We'll go step-by-step through the blog creation process, investigating how they work and how they can help your business. Then, you'll create and publish your first blog and begin experimenting with how to use this technology to build trust and visibility with your clients.

You'll learn:

- What blogs are, how they work, and where to find blogs related to your business and interests.
- How blogs help businesses and how they differ from Web sites, newsletters, and discussion forums.
- How to create and publish your first blog.
- Advanced blogging techniques for further control and flexibility in communicating your message.
- How to promote your blog and use it to reach out to prospective clients and communicate with them in a voice that is personal and friendly.

So let's get started!

PART ONE

A WHO-LOG? A WEB-WHAT?

In part one, we'll cover:

- The nature of blogs: what they look like and how they work.
- Where to find blogs to read.
- A brief history of blogs.

What Is a Blog?

For the elegant simplicity and beauty that blogs create, they certainly have an ugly, ungainly name. *Blawg* sounds like some kind of gurgling sound your cat makes prior to coughing up a giant hairball.

A blog, short for *Web log*, is often described as an online diary. The most recent entries (called *posts*) appear on the home page of the blog with links to archives of older posts. Archives are organized by date and can be categorized by topic. Often, each post will have a form for readers to add their own comments and to give opinions or reactions to the post content.

But that's just what a blog looks like. It doesn't really tell you *why* a blog is different from a conventional Web site.

The real genius in blogs shows itself in *how* they

are written. You can update your blog instantly from any Internet connection on any computer anywhere in the world—even from your summer home in the Bahamas as you sip an ice-cold island cocktail. Blogs are designed for people who don't want to learn (or don't have the time to learn) HTML or Web design. You use a special type of blogging software (called a *blog platform*) to create and update your blog. You type your post into a simple online form, click "Publish," and it's instantly available online for the whole world to see. It's super easy. If you can send an e-mail, you can publish a blog.

So here is my updated description of a blog: A blog is an easily, instantly, and frequently updated Web site, focused around a topic, industry, or personality.

Where Do I Find Blogs to Read?

Knowing the definition of a blog is one thing, but the easiest way to get acquainted with the conventions of the blog format is to start reading them. The more blogs you read, the more you'll get a feel for the depth and breadth of style and subject matter they have to offer. Here are a few sources:

Blog Ecosystems

The collective hive of blogs is called the *blogosphere.* Links to high-visibility subjects and breaking news bubble up in the blogosphere first, before they ever reach your evening television news program. Blog *ecosystems* are sites that monitor those millions of blogs every minute of every day, searching for the top key words being mentioned in blog posts or the most popular links, books, movies, and people. Visiting these sites is like being in a huge convention hall eavesdropping on millions of conversations at the same time. This is a fantastic way to do market research and to see what everybody is buzzing about. Following is a list of several blog ecosystems:

Blogdex (www.blogdex.net)
IceRocket (www.icerocket.com)
BlogPulse (www.blogpulse.com)
Technorati (www.technorati.com)

Blog Search Engines and Directories

Just as there are for Web sites, there are also directories devoted to cataloging the ever-growing global network of blogs. Try searching by topic or geographic location.

Blogwise (www.blogwise.com)
Blogarama (www.blogarama.com)
Globe of Blogs (www.globeofblogs.com)

Search Engines

The popular sites below are the more traditional, more widely known search engines, and are a great place to start searching for blogs. You can type in a topic followed by the word "blogs" and see what results come up. The more specific the key word, the more likely you are to find bloggers talking about what you're interested in.

Google (www.google.com)
Yahoo! (www.yahoo.com)
A9 (www.a9.com)

Blogrolls

Most blogs have a *blogroll*, which is a list of that blogger's favorite blogs and Web sites. If you find a blog you like, try checking out other Web sites that particular blogger enjoys. You might find other valuable, complementary sites and blogs to read.

Anatomy of a Blog

Now that you know what blogs are and have started reading them, you'll notice some common elements. Let's take a quick look at eight of the most common elements of a blog:

1. Title and Tag line

The more obvious your blog's title, the more quickly readers can figure out your blog's topic. Sure, a clever title is nice, but if you want to get the right kind of targeted traffic to your blog, be obvious. I'd rather see a great blog with a boring, obvious title (and fantastic content) than a clever title that no one can figure out. Also know that you don't have to have the word "blog" in the title of your blog. Include a tag line under the title to help add a little more clarity and focus to the topic of your blog. When you make your blog available to search engines, remember that your title and tag line should reflect important key words. What terms, queries, and solutions might your customers be looking for that relate to your products and services?

2. Posts

A blog's entries are called posts. Typically, the most recent posts are displayed on the front page, in reverse chronological order. As new posts are added, the older ones roll off the front page into the archives.

Posts can consist of quick write-ups of helpful links you've found online, opinion pieces, long essays, short thoughts—anything goes. There's no wrong way to write a post, except to be sterile and boring. If I want to be boring, I can go write a press release!

3. Permalinks

You'll notice many blog posts have a *permalink* at the end. Permalinks serve as the permanent home (a "permanent link") of a post, so others can link to it for future reference.

Let's say you link back to the latest post on my blog—which at the time is on my front page. As time goes by, the post you referred to moves off my front page and into my blog's *archives*. If you just merely linked to my blog's home page, anyone clicking on your link to my blog would end up at the front page and need to search the archives for what you were trying to link to. However, if you link to the specific blog post using the permalink, then your readers will be able to navigate directly to that post.

Think of it like this: Every post on your blog has its very own individual Web page. Instead of saying, "Go to the home page of Andy's blog, go into his archives from November, then go to the 4th, and it's the fifth post he wrote that day," the permalink allows you to link directly to that post.

Permalinks are a great idea—not just for blogs—but for all Web sites. To find a blog post's permalink, you typically click the post's title or click the word "permalink" located at the bottom of the post. You'll then be taken to an individual page for just that particular post. Your browser's address bar will reflect the URL of that post's permanent link.

4. Comments

A blog is made of posts, and, if you allow for it, readers can leave their *comments* right below each post. One of the biggest reasons blogs have taken off like they have is the ease of engaging readers in an immediate conversation. With a newsletter, you've got people e-mailing directly to your inbox. It becomes merely a dialogue between you and them—they don't see what everybody else is saying. With a blog, they can add their thoughts about your post using a comment form. It's like having a mini guest book for each post.

They can agree with, expand on, expound or rebut your post. They can even offer suggestions about where to go for further information. Allowing comments on your blog is a sign that you're open to discussion or collaboration on your topic. And, as the blogger, you can always remove, edit, or revise comments if necessary—or even turn off the commenting feature for individual posts. Most blog software also allows you to approve any comments before they show up on your blog.

5. Archives

As posts age, they roll off the home page and into the archives. This is where blog software really shows its moxie. Instead of you (or your poor Web designer) having to update dozens of pages every time you add a new post, the blog platform does it for you. It adds the new post to the top of the home page, moves the oldest post off, and rearranges the archives as needed. Archives are important because they are organized for easy indexing by search engines. Archives are typically organized by month or week, depending on what you specify.

6. Author Info

A blog with no author information is pretty darn boring. But a blog with no *interesting* author information is worse. The more individual and candid the details, the better. A blogger who tells not just what he does but who he is builds a stronger relationship with readers than just being Regional Director of Something Important-Sounding. I used to post my chili recipe on the front page of my blog and people would remember my site because of it—"Oh, you're the guy with the chili recipe."[2]

---

[2]Andy's Chili Recipe: Brown 1 lb ground turkey. Add 2 cans crushed tomatoes and 1 packet of chili seasoning. Chop/add 1 onion, 2 green peppers, 1 clove of garlic, 1 bay leaf, and 6 stalks celery. Toss in beans if desired. Add cinnamon to taste. Cook all day. Serve over pasta or rice. Add in chopped cooked Italian sausage for extra meaty goodness.

7. Blogrolls

As mentioned above, a blogroll is a list of your favorite blogs and Web sites. Just like the author information, it tells the reader a little bit more about who you are. Blogrolls contribute to the networking of blogs and the cross-linking of ideas.

8. Feeds

You may see a link that says "Subscribe to this blog's feed." Or a big orange button that says XML or RSS. These are links to a blog's *feeds*. Feeds are another way for your readers to get the latest updates on your blog using a type of software called a news aggregator. Feeds are becoming popular because they provide readers a spam-proof, virus-free way to receive updates like newsletters. As feeds are a pretty complex topic, I've dedicated a whole chapter to this subject, on page 140.

SUCCESS STORY

THE NEW PUBLIC RELATIONS

Marketing consultants Scott Allen and David Teten have been using their blog to help promote their book, *The Virtual Handshake*:

I think the most dramatic way in which blogging changed our business, and in which it can change other businesses, is that it has become the focal point of both our content and our connections. Anything that we read or write of any substance that's related to our topic ends up on the blog, making it an invaluable archival resource for anyone interested in the topic, and especially for us. E-mail is still the killer application for keeping up with people virtually, but blogging is wonderfully powerful in this regard because it's public. It's one thing to send somebody a note of congratulations—it's a whole other level to post about it in your blog. With the e-mail, the benefit to them is purely emotional. With a blog post, the benefit is very real, albeit virtual. It publicly portrays them in a good light, plus gives them a link, which helps their search engine rankings. And, of course, as Phil Agar said, the most fundamental way to find people on the Internet is to make it easy for them to find you. Our blog makes us highly visible on the Internet for our topic and has attracted media, customers, and potential business partners as a result.

You can read Scott and David's blog at www.thevirtualhandshake.com.

What Is a TrackBack?

Often you'll see a link at the bottom of a post that says *TrackBack*. A TrackBack links to another blog that is referencing that particular post. Here's an example:

Jane writes a post on her blog about "Online Business Basics."

Bob reads Jane's post and publishes a post on his blog that links to hers.

When Bob publishes his post, his blog platform goes and talks to Jane's blog platform and says, "Hey, we're talking about Jane's post over here on Bob's blog."

Jane's blog says, "Thanks, I'll post the reference." To the bottom of Jane's post the blog now adds a link—a TrackBack—to Bob's post talking about "Online Business Basics."

So now, along with the comments of a post, there can also be a list of other blogs referencing the post. Jane's blog is telling you which other sites and bloggers are talking about her post.

I've often heard a TrackBack described as a "remote comment." Instead of writing your comments at the end of a post, you write them on your blog. The process of creating the links between the blogs is automatic when TrackBack is activated.

TrackBack is another feature (like permalinks) that greatly enhances the connectedness of Web pages and the overall Internet.

How a Blog Works

Now that we've laid the groundwork, let's take a closer look at exactly how a blog works. The example below illustrates how the blog platform TypePad can be used to publish your blog.

1. Log in to www.typepad.com from any Internet connection on any computer anywhere in the world.
2. Fill out a simple Web form with your latest post and click publish.
3. TypePad does the rest. It adds this new post to your database of posts. It then grabs the posts and pours them into the blog's template, generating the blog's structure.
4. TypePad then publishes those blog pages on the Internet.

5. Your updates are available online, instantly, for anyone in the world to read.

Let's review what happened: Log in, write, post, publish. And you did all this without having to contract with a Web designer or contact an IT department.

Some blog platforms, like TypePad, provide the platform and the hosting for the blog. Others allow you to send the blog's files over to be hosted by your own server. There are blog platforms that can be installed on your own Web site's server to give you complete control of the look and feel of your blog. In part three, we'll talk about factors that go into picking a blog platform.

To view an animated demo of this process go to http://www.goblogwild.com/goodies/.

A Brief History of Blogging

So where did all of these blogs come from? How did they evolve from an underground geek toy into a mainstream tool that has revolutionized politics, journalism, marketing, and the media?

I always find that when I learn about a new technology or way of doing something online, the geeks have already been there. (And I say the word *geek* with complete affection. I'm one too!) Before business became inseparable from the computer and the Internet, the academics and tinkerers were there first—kicking the tires and playing with the technology. Blogging was no exception.

Many of the earliest Internet geeks authored online journals, expressing their viewpoints and adding links to other interesting things, but because this was before the availability of software like FrontPage and Dreamweaver, everything was done by hand—

and that was a huge hassle. Each time they updated a site, they ran the risk of breaking links inside the Web site and possibly leaving readers stranded with the dreaded "page not found" error. Eventually, computer programmers and Web designers became tired of this manual work and created their own software to automate the process of updating their Web sites. Thus the blog was born.

In 1999, a Web site called Blogger (www.blogger.com) launched, offering free blogs to anybody who wanted one. Blogger provided both the blog platform and hosting space for blogs. Millions of people from all over the world logged on and started creating blogs, and the world's largest conversation began. Blogs have changed the way academics do research, journalists write, families connect, and politicians raise funds. It was only a matter of time before businesses woke up and realized the power of blogging for marketing online. Blogging provides a way for companies and customers to meet on common ground to talk about what excites them and what makes them tick. It also allows them to closely track where, and under what circumstances, their products are being talked about online.

Today, blogging continues to grow at a swift pace. More than 70,000 new blogs are created each day.

More than a million posts are added every single day. Millions of people look to blogs for a good laugh, a great idea, a fantastic tip, or an instant analysis. The blogosphere doubles in size every five months, allowing anyone, anywhere, in any language, to make their mark on the world—to share their ideas, passions, and products.

According to *Merriam-Webster's Collegiate Dictionary,* "blog" was the word of the year in 2004, the year that marked a turning point for blogs as one of the most sought-after media formats. At that time, blogs were seen as a venue for gathering large audiences of devoted readers who used blogs to supplement, complement, and fact-check their nightly news.

My favorite milestone for blogging was when "Blogs" appeared as a category on the *Jeopardy!* game show. *I'll take blogging for one thousand, please, Alex!*

The Rise of Citizen Journalism

Blogging also points to an emergent trend in technology—the rise of *citizen journalism* (this is also called "consumer-generated media"). Blogging supports the idea that customers and consumers are not passive receptors of media and marketing, but are themselves active participants in the media. As technology for instant self-publishing is becoming cheap, easy, and ubiquitous, anyone can say anything. But, of course, the drawback is that anyone can say anything.

Often, blogs are the first source of information for breaking events, as television stations rush to get vans on the scene and newspapers wait for the next edition. Bloggers are able to publish text, pictures, audio, and video on the spot, instantly. For businesses, this means looking at blogs as an ongoing customer survey. Perhaps your company's blogging approach won't be publishing a blog, but reading the ones that are already out there. Blogs drive transparency—whether a company likes it or not. (Remember that pesky Dan Rather memo?)

Finally, blogs create a greater sense of trust and reputation than traditional media. Just as customers trust advice from friends, readers trust recommendations from bloggers more highly than from traditional advertising. Barriers for consumers to get online and vent (or praise) have all but disappeared. Now even Grandma can wreck your ad campaign.

Who Reads Blogs?

Advertising network BlogAds (www.blogads.com) conducts an annual survey of thousands of blog readers, comparing their habits to the general online audience. The results are surprising.

- Blog readers are older, with more than half over the age of thirty.
- Blog readers have money to burn. Seventy-five percent make more than $45,000 per year.
- Blog readers shop online for everything from health foods to political contributions. They are Web-savvy and not fazed by online shopping carts or transactions.
- Blog readers have one thing that is lacking in many other public arenas: passion and initiative.

Blog readers actively seek out the latest and greatest information and products available online. They click ads to find out more and to make informed buying decisions. They are influential opinion-makers—the kind of folks that others turn to for the latest news on what's hot and what's worth buying. Blog readers see blog writers as the experts—as movers and shakers that are working on, or searching for, the next big thing.

• • •

By now, you've got a good understanding of what blogs are, what they look like, where they came from, who reads them, and how they work. You've also started to find blogs that might relate to your particular business or industry. Be sure to bookmark blogs you find compelling or interesting to revisit for further inspiration and ideas.

PART TWO

BLOGS AND BUSINESS

Now that you have a handle on what blogs are and how they work, we'll check out what that means to business and look at:

- How companies use blogs to market and promote their businesses.
- Examples of companies using blogs in concert with Web sites.
- Comparisons of blogs to discussion forums and e-mail newsletters.
- How blogs make money—often thousands of dollars—for bloggers.

Let's go . . .

Blogs and Business

Now that you have a basic understanding of the ins and outs of blogs, and of some of the options available, you're probably wondering "How can blogs help *my* business? How do they help *me* to stand out?"

Blogger and Web designer Angie McKaig (www.angiemckaig.com) nailed it when she outlined these three ways that blogs can enrich any business:

- They offer fresh content on a daily, or at least on a very regular, basis.
- They present an informal voice that visitors can respond to and get to know.
- They provide useful information via links.

These three points are the foundation of how blogs can help companies. Here are some other ways that blogs can be used for marketing and building a business.

Communicate with Your Team. Use an internal blog to communicate project status to stakeholders and managers. It beats clogging everyone's e-mail with multiple mailings, and it allows these missives to be archived, indexed, and easily searched. Blogs are perfect for knowledge management.

Enrich Your Clients' Understanding of Your Business. You can easily link to white papers, articles, and resources relevant to your readers and their needs. You can more easily attract experts to provide value-added content to your audience by hosting interviews or Webinars to educate clients about trends affecting their industry.

Reach Out to Your Customer. Nobody buys from someone they don't know. Blogging allows you to demonstrate your expertise and point of view quickly and easily. In addition, blogs allow customers to receive your updates in the format they choose: e-mail alert, online, or through feed technology (more on feeds later). You can even launch a "faux blog" and have your company's mascot or branded character report their experiences working with your company.

Build a Buzz. You can create your own marketing buzz to drive attention and buyers to your products and services. Courting influential bloggers in your industry to review your latest product or service is a great way to generate attention before a product launch. Companies like BzzAgent (www.bzzagent.com) and Marqui (www.marqui.com) have used bloggers as prime movers in their clients' marketing campaigns.

Test Drive New Ideas. Blogs are the perfect forum to test out new ideas and receive instant feedback. General Motors does this with their Fastlane Blog (fastlane.gmblogs.com). You can allow others to see how you develop your products and services, and at the same time, they can tell you how best to serve them.

Go Global. Blogs, like other online media, allow you to take your business and ideas to a global market. Translation services are getting better every day, allowing more people to read online content in various languages. I've helped bloggers from New York to New Zealand, from Indiana to India.

Create a Backdoor to the Press. Journalists are busier than ever, and blogs provide a virtual directory of pundits on any topic. You and your company might be the content experts they're looking for. Furthermore, if your company is talked about in the blogosphere, it could lead to exposure in the mainstream press.

Write Your Book. Blogs can not only give publishers a taste of your writing style, but also allow them to assess your level of expertise. Let your readers help you write your next book or article. Post chapters or ideas, then let readers help you in researching, testing, and suggesting ideas. Or use a blog after your book is published to update the material or to answer questions from readers.

Highlight Success Stories. Invite clients to blog about their successes with your products and services—it's like creating a living testimonial that never ends. As clients share their experiences, your prospective clients can see, firsthand, how you can help them, too.

Communicate in a Crisis. When companies are in a PR pinch, need to get accurate, timely information to the press, and don't have time to wait for the IT department to work through the tiresome change-control process, a blog might be just what is needed. A blog provides an instant way to get updates out (internally or externally) to the people that need them most.

For a more extensive discussion of blogs and business, check out Business Blog Basics (www.businessblogbasics.com).

Fire Your Web Designer

One reason businesses, especially small businesses and solo professionals, flock to blogs is that they reduce the dependency on a Web design team to get ideas and announcements online.

Developing a Web site for your business can take weeks, sometimes months, from shopping for a designer and negotiating a solution to finally getting the site launched. Developing a Web site can not only be costly, but also time-consuming.

Using a blog, you can create a Web site in less than fifteen minutes, for less than ten dollars, with hosting and software included. All this, without having to know anything about Web design. With a blog there are far fewer decisions to make, and you can be online and publishing in the time it takes to watch a sitcom.

Many of my clients have found that blogs are a

simple and inexpensive way to start out online while building their businesses, with a goal of eventually using more robust Web design solutions. I work with others who love the blog format so much they have abandoned their Web sites for it, vowing to never look back.

Become the Filter, Be the Lens

I was leading a blogging seminar one afternoon when a participant heckled: "Who reads these things? There's too much stuff online! I don't have time to do all of this reading!" Neither do your prospects, clients, and colleagues.

Think of all the information that your potential customers and current clients need to know to succeed in their businesses and in their lives. Without a doubt, they don't have time to read a hundred marketing magazines, journals, and Web sites each month to get the information they want. Instead, they can just find a handful of bloggers to find, field, and filter the information for them.

What if you can become the filter for your customers? You can provide a digest version of the latest industry trends and news that matter most to them.

You can become the blog they can't live without. Over time, you'll build a reputation with them. As their trust in you grows, so will their inclination to purchase your products and services. Think of a blog as a sales brochure that never ends.

Web Site or Blog?

You're probably wondering if you should have a Web site or a blog, or both. There is no right answer, but here are some things to consider:

- Revisions to a full-scale Web site typically cost a lot more to set up and update than a blog.
- A company's traditional Web site might share a different point of view or "voice" than its blog, and keeping them separate might make that difference clearer to the readers. A blog's style is more conversational, while a Web site is more formal.
- A Web site typically has a longer approval process for changes and revisions, and often comes with a busy IT department or Web design studio to negotiate with. With a blog, the level of control is reduced to the authors of the blog.

Let's take a look at what two different businesses are doing.

Separate Blog, Separate Web Site: Seth Godin

Marketing guru Seth Godin, author of books like *Purple Cow* and *Permission Marketing*, went the route of having a separate blog and Web site. His Web site (www.sethgodin.com) contains the usual ingredients of a business Web site: who he is, what he does, his books, and upcoming events. Overall, the tone is friendly, but traditional. His blog (sethgodin.typepad.com), on the other hand, is a little more freewheeling. Here you can read his day-to-day adventures as Seth researches his next big marketing idea. It is a peek under the hood of one of the top marketing minds working today.

Web Site and Blog Hybrid: The Tom Peters Company

Everybody loves Tom Peters, one of the brains behind *In Search of Excellence,* for his raucous brand of business management evangelism. His company's Web site (www.tompeters.com) effectively blends the best of a Web site and a blog.

On the left column of his blog, you'll find the traditional elements of a Web site, the static pages that don't change very often. In the center column, you can view a running commentary titled Dispatches from the New World of Work that is written by Peters and his staff. This includes resources and articles from business magazines and external Web sites.

What if you could get inside the head of one of the top management gurus in the world? Now you can.

Depending on your budget, schedule, and business needs, you will want to consider if a blog will best benefit your business as a separate entity or integrated into your existing Web presence.

The Legend of Microsoft

Blogs can also be used by large corporations to soften and humanize their image. In this case, Microsoft allows employees to blog about their day-to-day experience working for one of the biggest tech companies in the world (blogs.mdsn.com).

Each of their bloggers presents a more human face of the software giant and allows us to get past seeing Microsoft as an impersonal, monolithic software company (or that annoying paper clip).

Check out Robert Scoble's blog (scobleizer.wordpress.com) for an example of one of their employees blogging openly, honestly, and passionately about working for one of the software giants. You can't force or manufacture this kind of trust—this isn't about a one-way brand, but about a real personality. In counterpoint, there are blogs like Mini-Microsoft (minimsft.blogspot.com), a blog written by an anonymous employee of the company, who is openly critical and challenges the company to fix faulty business practices and cut the fat out of the organization.

What if you allowed your employees to blog about their experiences working in your company? Are you open to that criticism? What if the marketing and legal departments didn't get in the way? What if you talked to your customers like you talk to your friends?

SUCCESS STORY

GENERATING PRODUCT AWARENESS

John Nardini works for Denali Flavors, an ice cream company with specialty flavors like Moose Tracks, Caramel Caribou, and Bear Claw:

One of our business objectives was to generate awareness of the Moose Tracks brand, which would then lead to a trial purchase. Once customers took the plunge, our experience showed that the product's taste would drive repeat business. We decided to accomplish this objective by creating a series of blogs aimed at different consumer groups. The blogs would link to the Moose Tracks Web site and be designed to funnel visitors to the site. In this way, awareness of the product would happen naturally. In addition, the advertising/promotion costs would be low compared to a traditional media effort. As each site was developed, it was marketed using guerrilla marketing tactics such as posting comments on other blogs (with links back to the original blog), trading links with other sites, asking other sites for referrals, and writing articles on sites that would link back to the blog. All of these efforts drove traffic to the blogs, which, in turn, drove traffic to the Moose Tracks site. The initial results have been very positive with higher site traffic and more visitors spending more time on the site. In addition, the costs are very low. Other than the time of our blogger, the company has spent less than $500.

You can visit the Denali blogs at www.moosetopia.com and freemoneyfinance.com.

E-mail Newsletter or Blog?

I'm often asked if it is better to use an e-mail newsletter or a blog. My answer is always the same: Use both!

A blog makes managing your past newsletter archives easier, brings newsletter readers to your site, and centralizes responses and feedback. Instead of having to mess around with getting each edition of your newsletter online, you simply add the newsletter to your blog and it's online immediately and archived for posterity. Also, if your newsletters are archived, they can be indexed more easily by search engines, providing you and your newsletter with increased visibility.

Instead of asking newsletter readers to e-mail you their feedback, send them to your blog to post comments. You won't be flooded with e-mails, and readers can go online to read what everybody else is writing.

A more sophisticated use of this technology is to have your blog create your newsletter for you. You can take your blog posts, combine them with readers' comments, place this content in your e-mail newsletter template, and then e-mail the finished product. It's that simple.

Blogs vs. Discussion Forums

Another common question I'm often asked is how is a blog different from a discussion forum, message board, or e-mail–based discussion list? A post on the technology forum TechSoup (www.techsoup.com) describes the differences:

> *Blogs are diaries that can be read by the public, while bulletin boards are town hall meetings in which the public can all discuss issues equally. The town hall quality of message boards can play off of the informational/journal quality of blogs.*

Visit any blog and you'll experience the running commentary and thoughts of one person, or of a small group of people, where the potential exists for the blog to be read or broadcast to a much larger group.

New topics are started by the blog's authors, and anyone can comment on the posts once a topic is posted. However, in discussion forums, you'll see more of a content free-for-all. Anyone can start a new topic or even branch off existing topics into more specific, targeted discussions. Furthermore, a blog doesn't depend on interaction and commenting. The blog functions without the input of others, with the notable exception of the author. A discussion forum, by its very nature, requires interaction and commenting.

How to Make Money Blogging

So where's the money in blogging?

While blogs can help you create credibility and improve relationships with your clients, there are also a number of ways to make money directly from your blog.

Sell your own products. Use a blog to talk about how others are using your products. Allow comments from readers, and provide space where clients can post testimonials for others to see. You can create products easily with services like CafePress (www.cafepress.com) and Lulu (www.lulu.com).

Develop a "members only" blog that requires a membership fee. Use a public blog to create interest and then a private one for exclusive access to you and your ideas.

Get a sponsor. Once you've created a dedicated audience, you can find a sponsor that needs to get in touch with that audience. Have them underwrite your blog expenses. Companies like Volvo, Nike, and Adobe have done this to increase brand awareness and to promote their products.

Include affiliate sales opportunities. Writing a book review? Include your Amazon affiliate link and suggest related products. Tease the reader with how a product has changed the way you do business and then provide a link so they can buy it immediately.

Use hosted ads. Services like BlogAds (www.blogads.com) or Google AdSense (www.google.com/adsense) will place ads on your blog that are related to your content based on key words. PVRblog (www.pvrblog.com), which covers news about personal television recorders like Tivo, pays for its hosting using this method and was one of the early examples of this income stream.

Launch a blog network. Companies like Weblogs, Inc. (www.weblogsinc.com) and Gawker Media (www.gawker.com) have blazed the trail in creating entire networks of blogs staffed by expert writers. These sites earn revenue through ads and affiliate sales on related products. Other blog networks include b5media (www.b5media.com) and Manolo's Shoe Blog (www.shoeblogs.com).

Add a tip jar. Ask readers to send you some dough through PayPal or Amazon if they find your blog compelling or useful.

Bloggers are making thousands of dollars a month with these strategies.

For a complete guide to making blogging an occupation through building your own blogging empire, check out Six Figure Blogging (www.sixfigureblogging.com).

Why Google Loves Blogs

A while back, I ordered used books through a reseller on Amazon.com's Marketplace. After six weeks, I still hadn't received my books. I knew that Amazon was just the matchmaker for the sale, so they probably wouldn't do anything. After e-mailing multiple times and receiving no response, I wrote a post about the guy that ripped me off, using his name as the title of the post. Within forty-eight hours, eight other jilted customers had found my post and had left similar comments and commiseration. I checked the Google search engine, and, sure enough, I was the number one result for those searching for this individual—even above his own Web site!

Here's why Google (and other search engines) like blogs so much, and why it gives them more attention (and got me more attention!) than conventional Web pages:

- **Frequent updates.** Search engines often weight fresh content with greater importance. Blogs are updated much more frequently than conventional Web sites.
- **Linked and networked.** Blogs are likely to have more outgoing and incoming links than traditional Web sites. These links significantly effect page rankings.
- **Built for indexing.** The underlying HTML behind blogs is typically much simpler and cleaner, allowing them to be more easily and accurately indexed by search engines.
- **Archived and organized.** Blogs have better "architecture," which makes them more readable to search engines.

My Amazon experience had a happy ending. The reseller e-mailed me his sincere apologies along with the books I had paid for months earlier. I modified my blog posts to reflect his follow-through. Moral of the story: Blogs are powerful!

Because they have such a strong and natural affinity with search engines, blogs can promote you to prominence in record time. Of course, that can be good or bad depending on the circumstances.

You Don't Need to Blog Every Day to Be Successful

Finding time to blog is one of the biggest obstacles preventing potential bloggers from getting started. Since most prominent bloggers blog every day, sometimes multiple times per day, many beginning bloggers think they need to keep pace. Lies, I tell ya! You don't have to blog every day to have a successful blog.

I recommend beginning bloggers **commit to writing at least three posts each week.** Surely you can come across three things worth telling your readers over the course of seven days. If you can't, then I'd say you aren't on the pulse of your target audience or industry, which is an entirely different matter altogether.

The e-mails you write to friends and colleagues, the interesting and enjoyable quips, ideas, or comments you come across in your daily life, and even the ideas in articles and books you've read—all of this can become content for your blog.

SUCCESS STORY

KNOWLEDGE MANAGEMENT

Online business manager Tina Forsyth helps entrepreneurs transition their off-line businesses to online businesses:

I consider my blog to be the hub of my business . . . I started my " 'Sneak a Peek' at what we are up to . . ." blog just over a year ago. I didn't have big plans for it at first, I just wanted a place to start capturing what I know about building business online. Recently, I took a look back at my posts and realized I've accumulated a gold mine of information. My blog literally contains the blueprint of how to run an online business. I even refer to it myself when looking for tools or tips. I plan on turning my blog into a published book in the next year, something I would never have considered if I had not had the blog to capture what I know. So in my humble opinion, everyone should have a blog for this reason alone—as a place to capture what they know, the "meat" of their business, their learnings.

You can read Tina's blog at www.onlinebusinessmanager.com.

Hot Topics, Hot Traffic

If you research sites that monitor blogs and their popularity, you'll find that the most popular blogs are either technology- or politics-based. The prominence of tech blogs is easily explained: People who read self-published Web sites about technology are more likely to use this same technology to publish their own views.

The prevalence of political blogs among the top hundred blogs is for two main reasons:

- Blogs depend on a constant stream of new content (via a twenty-four-hour news cycle powered by cable news, network TV, and print newspapers and magazines).
- Blogs offer a distinct, honest, passionate, and provocative perspective on politics.

Political blogs don't have to clamor for content because there's a huge media machine providing a never-ending channel of topics for debate. Political content has its own built-in passionate audience.

The success of technology- and politics-based blogs can guide you in your blogging endeavors. Keep these questions in mind when planning your business blog:

- Where is the twenty-four-hour news cycle in your industry or profession?
- How can you take current events and developments in your industry and put your own personal stamp on them?
- How can you take events from other industries or fields and apply them to your business?
- What are the dirty little secrets of your industry—those "elephants in the room"—nobody likes to talk about?

By focusing on these key areas, a blog can really shine. Be controversial. Be passionate. I often tell clients, "If nobody is commenting on your blog, you're not being honest enough." You don't have to be provocative just for the sake of being provocative—

After the Storm

Eventually the hype surrounding blogging will wear off, and everybody will move on to the next bright, shiny, new thing. What will remain is the simple, elegant usefulness that blogs have brought to Web publishing.

Instant Publishing

In the future, we'll all say, "Well of course you can edit any page on your site easily and instantly—what do you think this is, print?!" I find beginning bloggers are totally blown away by the immediacy of blogging. They just type a few paragraphs and hit Publish, and instantly their thoughts are available for all to read. As one of my clients commented, "It's like heroin for writers."

Globalize and Mobilize

As digital tools and formats converge, blogs become more useful as a way to catalog life and business with writings, files, links, pictures, audio, and video—posted from anywhere in the world. Services like Audblog (www.audblog.com) help you publish audio and video to your blog, even from a cell phone. This form of mobile, on-the-go blogging is sometimes referred to as *moblogging*.

Knowledge Management for Individuals

In the corporate sector, knowledge management means taking what's in the brains of the workforce and organizing it into a searchable, accessible database. Blogs are knowledge management for individuals. Spend a year blogging on your topic, and I'll bet that at the end of that year, you'll have enough content, concepts, and case studies to write a book in your chosen field.

Multiple Formats

Blog tools allow you to distribute your content in multiple formats: e-mail newsletter, Web page, handheld—

even formats that are easily read by hearing- or sight-impaired users. This flexibility increases the usefulness and accessibility of online content for everyone. With more blog platforms supporting multiple languages, content creation is increasing exponentially.

Separation of Content and Style

Because blogs keep your content in a database and then pour it into your templates, updating the look and feel of your Web site is infinitely easier. Most blog tools use Cascading Style Sheets which allow fonts, colors, and other elements of your blog to be controlled from one file, instead of weighing down each page with the same formatting information. This cuts bandwidth considerably and makes updating the design of your blog a matter of a few clicks of the mouse.

Are You Ready to Start Blogging?

By now, you have a pretty good handle on why you might start a blog to help market and grow your business. We discussed how blogs can help you build trust and a reputation in your field, methods for making direct income from blogs, and what might be next for this compelling online medium. In the remainder of this book, we'll focus on the how of blogging and walk through all of the simple steps you need to make to create and publish your first blog.

If you need a break, go get some coffee or tea. If not—let's get started!

PART THREE

CREATING AND PUBLISHING YOUR FIRST BLOG

Finally we get to the fun stuff! In part three, we'll walk through the process of creating and publishing your first blog.

You'll learn:

- What to consider in selecting a blog platform.
- How to set up your TypePad account and create a blog.
- How to publish and format blog posts.
- How to add pictures and upload files to enhance your blog.
- How to schedule blog posts to appear while you're on vacation.

So let's get going . . .

Choosing a Blog Platform

Once you've decided to start a blog, the next major decision is what blog platform to select. There are dozens of blog platforms to consider—each with its own strengths, weaknesses, and price points. Here are the basic considerations.

Blog Hosting: Do you want to host your blog on your own Web host's server, or a server maintained by the blog platform's company? If you have an IT department, company policies may dictate that you host your blog on a company-owned server for control and security purposes. If you don't have an IT department (or know an affordable geek), you may want to let another company handle your hosting. Some sites (like Blogger's Blog*Spot—www.blogspot.com) offer free hosting—but who are you going to complain to when your provider goes down?

Blog Platform Hosting: Some blog software is installed on your own server, while others are hosted by the company that developed the platform. Hosting with the platform provider gives you someone to call if you need technical support or assistance. Movable Type (www.sixapart.com) is software you host on your own server, which gives you more control over your blogs.

Multiple Blogs: Some blog software is configured to host multiple blogs, all through the same interface. This is a great feature for businesses that want to easily manage different blogs for different audiences. ExpressionEngine (www.expressionengine.com) is among the platforms that offers this feature.

Multiple Authors: Many platforms can have multiple authors so that you can give your team, clients, or anyone the ability to create posts. Blosxom (www.blosxom.com) is an example of a platform that allows multiple authors.

Multimedia Support: If you plan on managing photo albums or audio, you might consider a blog platform that supports these features. Nucleus (www.nucleuscms.org) is one of many that support image uploading, photo gallery generation, and audio blogging.

Comment Spam: Like any other online medium, blogs are a hot target for spammers who see a blog's comment forms as a free billboard to plug their wares. Programmers of blog platforms are constantly sharpening support for blacklist key words (like "Viagra" or "online casino") and other indicators that a comment may actually be an advertisement in disguise. Other safeguards include the fact that most platforms allow you to approve all comments before they are viewed by readers. Also, most blog platforms allow you to require users to register in order to leave a comment.

Multiple Formats: Will you need to output your blog in multiple formats? Some blogs may need to be read by handhelds, cell phones, or devices for sight- or hearing-impaired users.

Traffic and Statistics: Some platforms track where your blog's traffic is coming from so you can see what pages are most popular and what search engines and sites are sending you traffic. Services like Site Meter (www.sitemeter.com) allow you to track traffic by simply adding a small Site Meter graphic to your blog.

Template Engine: Different platforms have different levels of sophistication when it comes to customizing the look and feel of your blog. Some rely on your knowledge of computer languages like PHP or your familiarity with HTML and Cascading Style Sheet. Other platforms, like Textpattern (www.textpattern.com), use their own system of tags to allow users to shape the style of their blogs. These tags come with a learning curve, but offer greater flexibility.

Price: Blog platforms also come in a variety of prices and plans. Hosted plans like TypePad are billed monthly, while others like ExpressionEngine and Movable Type require you to buy a license. Other blog platforms, like the very popular WordPress (www.wordpress.org), are totally free under an open-source software license.

Head spinning yet? There are dozens of details to think about when choosing a blog platform. But I usually suggest beginning bloggers start with TypePad.

Why TypePad?

For beginners, I recommend starting out with the TypePad blog platform available from Six Apart. Let's look at some of the reasons why:

It's hosted for you. With TypePad, both your blog and the blog platform are hosted for you by Six Apart. You don't need to install anything or contract with a Web designer. Just register for an account and start blogging.

Multiple blogs and authors. If your blogging starts to grow, TypePad allows you to easily add new blogs and authors to your account. The Pro version of TypePad allows you to create an unlimited number of blogs.

Multimedia support. TypePad includes features for creating photo albums, uploading files, and easily adding images to your posts. The platform also integrates audio blogging services and is enabled for podcasting.

Comment spam control. Because TypePad hosts so many blogs, they can keep a tight watch on the same comment showing up on multiple blogs at one time, a sure sign of comment spamming. They also use a blacklist of key words to keep tabs on comments using common spam text. Once I had a spam comment on a TypePad blog and by the time I had logged in to remove it, the TypePad system admin had already purged the comment from the system.

Template engine. Tweaking the look and feel of your blog is easy with TypePad. Some bloggers find the template system constraining, but most are happy to have 10,000 less design decisions to make. Upgrading to the Pro version of the platform allows a great deal of latitude in the style of your blog. You can also use the advanced templates to design your blogs in multiple formats.

Traffic and stats analysis. Detailed traffic reports are provided by TypePad that can tell you where your blog's traffic is coming from, as well as which search engines are sending you surfers.

Easy come, easy go. Many blog platforms allow you to export your entries out of the system if you decide to change platforms or import your entries from another system. TypePad offers both features, so you're not tied down to using TypePad for the rest of your life.

Price. TypePad's service ranges from five to fifteen dollars per month. I recommend you start with the intermediate level plan, which has enough features to get you started. It's easy to upgrade if you find you need more options and flexibility.

It's easy. I've taught many clients how to use TypePad, and they consistently report back how simple the tool is to use. If an eighty-year-old ex-Amish man living in a retirement home can blog, so can you.

I feel that TypePad gives the most professional results out-of-the-gate. With other tools, you have to know a little more HTML or code to get results that are going to look impressive. The possibility of miscoding an HTML tag and throwing the whole design off can annoy users that don't want to learn coding. Most bloggers just want to write, publish, and be heard. They don't want to get caught up in the technical details.

I've included tips and strategies throughout that can be used with any blog platform; however, the tutorials in this book are based on the TypePad platform. If you decide to use another platform, reference its documentation and support forum for further guidance.

Crafting Your Blog's Persona

What are you going to talk about, and how are you going to talk about it? Those are the important questions we'll tackle in this section. We'll look at eight different aspects of a blog's persona. You're not putting anything in stone right now. It's just important to know all your options and to consider how different approaches and points of view can shape your blog. Go with your gut. It's better to make a decision now and start publishing your blog than to get mired in the details in an attempt to make everything perfect at first post. You can always change your mind.

Tone

Your blog can be super-personal (intimate revelations) or super-professional (just the facts, ma'am), or something in between those extremes.

Consider how your target audience will respond to that personal touch. It's a delicate balance to make your blog personal enough to be in an authentic human voice, while still keeping the focus on your business, profession, products, or services. If you're too candid, you'll alienate the more conservative side of your audience. If you're too sterile, why bother even having a blog?

Attitude

Blogging's frequency and format allow you great latitude in attitude. Some bloggers create a bad-boy (or evil twin) persona that helps them establish a distinct personality that is very different from their off line self. Others go the exact opposite route and are unabashedly truthful. Both approaches are anchored in honesty—one with the permission of a pseudonym, the other with the permission of authenticity. For a great example of a persona blog, check out www.rageboy.com. For an example of a totally candid blog check out the blog of comedienne Margaret Cho (www.margaretcho.com) and see how she uses a completely honest voice to talk about her business and performances.

Content

The two main traditions of blogging are *diary* and *linkfilter*. By *diary*, I mean the "inner workings" of your business and life: "Today I did this, or we did this." The diary format is important because it adds an element of freshness. However, while diaries can be interesting, following in the footsteps of erotic diarist Anaïs Nin is probably not the best approach for a budding start-up. The linkfilter approach hearkens back to the initial use of blogs: "I found this thing online and here's the link to it. I totally agree or totally disagree for X reasons. What do you think?" You can filter through the daily noise to present essential industry news and views, your opinion, and an invite for others to agree, rebut, expand, or extend.

Team

Most blogs are solo efforts from singular professionals designed to create a very intimate reader-to-author relationship of trust, built over weeks or months of reading. With these blogs, readers get to audition your expertise. Other blogs are a group effort composed of either like minds, a collective such as a project team or a family, or even total opposite points of view as in a point-counterpoint blog like the political blog WatchBlog (www.watchblog.com). Readers and their collaborators respond by creating community and sharing knowledge.

Things to consider:

- Will your blog have comments?
- Will other authors post or suggest links and entries?
- Who are some of your colleagues that you could invite to be guest bloggers?

Access

The more public a blog, the better. It allows you to link to, and, in turn, to be linked back. A public blog strengthens the connections between Web pages and knowledge that makes the Web worthwhile. If you are working on a private project, a password-protected blog may be in order. Private blogs also can be used to generate income via "members only" access.

Perspective

From what perspective are you going to blog? Are you an expert sharing your success stories and theories? Are you a novice taking a "student" role and cataloging your education in a certain topic? You don't have to be a recognized expert in a topic to blog about it—just stay open to learning and to feedback along the way.

The Only Way to Start Is to Start

If you have no idea what your style is, simply start blogging. Your style will emerge over time.

Blogging and Ethics

Our final stop before actually creating our blog is to take a quick look at the ethics of blogging.

The immediacy of blogs goes a long way toward providing a raw, unedited, from-the-horse's-mouth feel to a company's message. It's important for the blogger to realize that his or her role is as a liaison between the company and the online sphere (and as a scout for emerging trends affecting the business as well). A business blog should not be famous for its blogger's personality (unless you're a maverick like Howard Stern or Richard Branson), but for the effectiveness of the blogger in getting out the message about the company's products and services, as well as for fielding questions and comments between prospects, customers, and the organization.

Getting Fired for Your Blog

There have been several cases of bloggers getting fired from their jobs because of their blogs. Most of these cases revolved around possible breach of contract in regard to proprietary information, or lascivious content, or, simply, the blogger acting like an idiot. You should never blog your company's inner workings if it might be considered an advantage to competitors.

Admitting Mistakes

Most entrenched corporate cultures are reluctant to admit mistakes, especially in public. A company blogger can go a long way in terms of bringing an informal voice to a company's message, but will still need to ensure this voice is in alignment with the company's overall point of view. Some companies may see presenting "lessons learned" as an approach that leaves them too vulnerable in the marketplace. Still, others may see this approach as a way of leveling with the customer or educating them on how the company has forged its path.

Keep Them Separated

If you enjoy candid writing that may turn off a more traditional, conservative audience, I'd recommend keeping your professional and personal blogs separate. I currently write two blogs, my eponymous business blog (www.andywibbels.com), and Andymatic (www.andymatic.com), my personal blog. On my personal blog, I often discuss topics—like panic attacks, my cat, politics, and toffee recipes[3]—that might not be palatable to a more corporate audience. If someone is an avid reader of one of my professional blogs, they might share my sensibility and enjoy the more personal side—if not, that's okay too. It really is a matter of personal taste and of how your target audience would perceive knowing a little more about who

---

[3]Toffee recipe: 2c butter, 2c sugar, ¼ tsp salt, ½ lb chocolate chips, 1c ground almonds. Combine sugar, butter, and salt in a saucepan. Cook/stir at medium heat until butter melts. Continue to cook at 285° until mixture is deep amber. Pour into a foil-lined pan. While toffee is hot, sprinkle on chocolate chips. Set for 1½ minutes. Spread chocolate evenly with a spatula. Sprinkle on nuts and then press the nuts into the chocolate (put a plastic bag on your hand). Cool. Eat. Repeat.

you are as a person versus how you present yourself as a brand.

I stick with the idea that everybody hates a press release because it has no sense of humor—it is purely functional with no personality. Yes, products and profits are incredibly important, but isn't it a bit ridiculous to be so formal? Why not talk to your customers in the same voice that you use to talk to your colleagues or your friends? A little warmth can go a long way in reaching a prospect across the coolness of an online connection. Remind yourself that most everyone hates being marketed to, but who doesn't mind a helpful or entertaining conversation?

And, if you think having your boss find out about your blog is bad, remind me to tell you sometime about when my dad discovered my personal blog (it all ended up just fine, and now he's one of my top comment writers!).

For a complete overview of the legal and ethical issues involved with blogging, check out the Electronic Frontier Foundation's *Legal Guide for Bloggers* at www.eff.org/bloggers/lg/.

SUCCESS STORY

ADDING THAT PERSONAL TOUCH

Yvonne DiVita has a business that helps companies market more effectively to women online. She was reluctant to start a blog:

I was so certain I needed a newsletter to support my message on marketing to women who shop online; I just couldn't see the benefit of a blog! I was one of those people who got dragged kicking and screaming into the blogosphere. Since starting my blog in March of 2004, I've become a worldwide expert in my field: marketing to women who shop online. I've received many accolades from other professionals, and . . . the truly wondrous thing is that I get business from my blog! A blog is just so much more personal, and so much easier to manage, than a Web site or a newsletter. I for one will never turn back. Blogging is more than a "thin Web site"—it truly is a conversation with hundreds of interested people. My blog serves me well because I've learned the value of a niche market segment, of building a network of valuable connections with other professionals and with customers/clients. A blog lets people meet the real me—not just the business professional, but the real person behind the corporate image. Now, through my blog, I can become a trusted friend. And that makes all the difference in the world.

You can read Yvonne's blog at www.lipsticking.com.

Creating Your First Blog

Let's get going!

To create your TypePad account, you'll need a credit card so you can sign up for the free trial. You'll be able to cancel at any time.

1. Go to **www.typepad.com.**
2. On the TypePad home page, click the button for the Free Trial. The first part of the **Create an Account** form displays.
3. Enter a **Member Name** to use when you log in to the TypePad system. This name cannot be changed once you create your account. You'll also enter a **Password**. Be sure to make a note of your TypePad account **Member Name** and **Password**. Further down the page, you'll enter your name, e-mail address, and birth date.
4. Next, **choose a domain** for your TypePad account. TypePad blogs can be hosted at

yourname.typepad.com or *yourname*.blogs.com. Once you choose a domain, you won't want to change it. I recommend a generic domain that can encompass any blog that you might host with TypePad. The easiest solution is to use your first and last name, or your company name, like johnsmith.blogs.com, janedoe.type pad.com, or widgetsinternational.blogs.com. All of your blogs will hang off this domain like yourname.blogs.com/myblog or yourname .blogs.com/my_other_blog. Later, you can point a domain name you've registered to your TypePad blog. So that johnsmith.blogs.com could be accessed through johnsmith.com.

5. **Choose a membership level.** I recommend the middle choice because it offers more flexibility and options. You can upgrade or downgrade your plan during the trial if you want to try out the different membership levels and features.
6. On the **Sign Up** screen, enter **your billing information**.
7. Next, **choose a name for your blog.** Choose something obvious but evocative. My current favorite is E. Coli Blog (www.ecoliblog.com), a blog for a law firm representing victims of

food-borne illness. You can change your blog's name at any time.

8. Choose a **Layout Structure** for your blog. You can select the number of columns you would like as well as where you want them to be. You can change your layout at any time.
9. Select a **Design Style.** For now, choose one that is close to how you want your blog to look. You'll be able to customize it further later on.
10. Choose a **Weblog Privacy** option. If you are just testing out blogging and don't want your blog to be easily found, simply select that your blog not be publicized. You can always go back and change this option later.
11. Next, you'll confirm your registration and settings.
12. Your TypePad account is now created, you are logged in, and ready to get started.

To view an animated demo of this process, go to www.goblogwild.com/goodies/.

Getting to Know TypePad

Notice that TypePad uses a tabbed interface to guide you through the different sections of the site. The four top-level tabs are:

Weblogs: You'll do most of your work here. This is where you manage your blog's posts and design.

Photo Albums: You can upload and arrange your TypePad account's photo albums here.

TypeLists: TypeLists are lists of your favorite things. You can manage all of your TypeLists for all of your blogs from this tab.

Control Panel: This is where you configure overall settings for your TypePad account, like your profile and billing information.

Each of the four main tabs has subtabs underneath it, as well as a third level of options after that.

Publishing Your First Post

Now that you've created your TypePad account, let's walk through writing, formatting, and publishing your first post:

Log in to TypePad. You'll see a list of the blogs in your TypePad account and a link that says **POST** next to one. For now, you'll see just the one blog you've created. As you add more blogs, they'll appear here when you log in to TypePad.

Click the **POST** link. The **Compose a New Post** screen displays.

Basic Posting and Text Formatting

1. In the **Title** field, enter a title for your first post. Type **Hello World!**
2. **Choose a category** for this test post. You'll be

able to remove categories or add your own later on.

3. Under **Post Body**, enter several paragraphs of text.
4. Let's make some of the text bold. You'll notice a toolbar in the **Post Body** field that works just like a conventional word processor. Highlight some of the text that you want to make bold.
5. Click the **Bold** button located on the toolbar (it has a big bold **B** on it). The text now appears in bold.
6. Repeat this process with the **Italics** button (an italicized ***I***) to add italics and the **Underline** button (an underlined **U**) to practice underlining text. Underlined text on Web pages usually refers to a clickable link, so make sure you aren't confusing your readers.
7. TypePad allows you to preview your post before you publish it. Click the **Preview** button at the bottom of the screen.
8. A preview of your post's text is displayed with the option to either **Re-Edit** or **Save** the post.
9. Click **Re-Edit** this post. The post editing screen displays, allowing you to continue work on your post.

Change Text Size and Color

1. To resize text, highlight the text and then use the text size drop-down menu to change the size from "Normal" to a different size.
2. To change the color of text, highlight some of the text, then click the **Font Color** button, shown as the letter ***A*** with a color palette. A color palette appears.
3. Click the desired color to change the highlighted text.

Add a Link

1. Let's add a link to some text. Type some text that you want to add a link to. For example, you might type "Click here to go to CNN."
2. Highlight the text you want to add a link to and click the **Insert Link** button on the formatting toolbar—it looks like a chain link.
3. A pop-up box asks you for the link you want to add—in this case www.cnn.com/. Enter the link and click **OK**.
4. The text is now linked to the URL you specified. The text won't be clickable until you are viewing it on the published blog.

Add an E-mail Link

Now let's add a link that allows people to send you e-mail.

1. Type the text: *E-mail me.*
2. Highlight the text and click the Insert E-mail Link button which looks like an envelope.
3. A pop-up box appears. Type in the e-mail address you want to link to the text, for example, I would type "*contact@goblogwild.com.*"
4. Click **OK**. Your highlighted text is now linked to the e-mail address you specified.

Indenting and Bulleting Text

1. To indent text, select the text you want to indent and then click the Begin Quote button (it looks like a left-handed quotation mark). You can also use this feature to quote a passage of text in a post.
2. The text you selected is now indented.
3. To make lines of bulleted text, highlight the text and then select either the **Un-Ordered List** button (which looks like a bulleted list) or the

Ordered List button (which looks like a numbered list).
4. The text you've selected is now bulleted.

Saving a Post

1. Once you are finished writing a post, click the **Save** button. TypePad saves your post to its database and publishes the post.
2. Click the **View Weblog** link (in the upper right-hand corner). Your blog opens up in a second browser window.

Revising a Post

You can edit any post on your blog at any time. To access your previous posts go to **Weblogs** > **(Your Blog)** > **Post** > **List Posts**.

Saving a Post for Later

You can also save a post for later publishing by saving it as a draft. To save a post as an unpublished draft, select **Draft** from the Post Status drop-down menu when editing the post.

Renaming Your Blog

If you want to change the name of your blog, go to **Weblogs > (Your Blog) > Configure**. You can modify your blog's name and tagline.

Pasting Text from Other Web Sites or from Microsoft Word

Copying and pasting text directly from other Web sites or from Microsoft Word into the post editing screen may cause you major headaches, as coding styles clash with the way TypePad formats posts. Your best bet is to strip the formatting by copying and pasting it into a text editor (like Notepad or TextEdit), which will remove any of the text's formatting. Then copy and paste the unformatted text into TypePad. You'll be able to reformat the text once you've pasted it into TypePad.

Adding Images to Your Posts

Adding images to illustrate and fluff up your posts is easy with TypePad. The system handles the uploading, linking, and management of any image file for you. TypePad accepts images in GIF, JPG, or PNG format.

1. When editing a post, click the **Insert Image** button on the button toolbar. A pop-up window opens.
2. Click **Browse** to locate the image you want to upload on your hard drive.
3. Choose whether you want the default **Image Options** or to **Use Custom Settings**. Custom Settings include:

   Wrap Text: Lets you choose whether or not you want post text to wrap around the image, and on which side.

Create Thumbnail: For large images, TypePad can put a smaller "thumbnail" version of the image in your post, and then readers can click the thumbnail to see the full-size version.

Pop-up Window: Choose this option if you want the full-size version of the image to appear in a second browser window.

4. Click **Insert Image**. TypePad receives the image and then inserts it into the post you are editing.

Uploading, Linking, and Sharing Files

Sometimes, you may want to post documents such as PDF documents, Excel spreadsheets, or other files. TypePad makes it easy to upload files, link them to your posts, and allow others to download them for review or distribution, all in a few easy steps:

1. While editing a post, click the **Insert File** button. A pop-up window opens.
2. Click **Browse** to locate the file you want to upload on your hard drive.
3. Click **Upload** file. TypePad receives the file and then puts a link to the file in the post you were editing.

Note: It's considered good form to include the format and size of files posted online—(Example: White Paper:

Blogs and Public Relations [PDF, 20KB]). That way, surfers on dial-up connections can judge how long it will take them to download a file.

Schedule a Post for Future Publishing

TypePad allows you to schedule posts for future publishing so your blog can be updated even when you're on vacation. If you know you're going to have a particularly busy week, you can spend a Saturday afternoon writing posts that will automatically be posted at specified times during the following week.

How to schedule posts:

1. When you are finished writing a post, scroll to the bottom of the **Edit Post** screen.
2. On the bottom-left is a drop-down menu with your **Posting Status**. To schedule a post for the future, select **Publish On . . .** from the Posting Status drop-down menu. A pop-up window opens with a calendar and time of day.

3. Select a time of day and a date for your post to be published.
4. Click **Set Time**. The pop-up window closes and the Posting Status now displays a time and date (to edit it again, click the clock icon).
5. Click **Save**. Your post is now scheduled to appear on the day and at the time specified.

Backdating Posts

You can also backdate posts to appear backdated in your archives. For instance, if you just started blogging this week but have had a newsletter for three years, you can post your past issues and backdate them so they show up in your archives.

• • •

By now, you're pretty familiar with the basic functionality of TypePad. We've covered what to consider in choosing a blog platform, how to integrate images and files into your blog's posts, and how to automate your blog publishing for ultimate flexibility. Now we will explore some more advanced techniques that will allow you to customize your blog to suit your business or personal needs.

PART FOUR

ADVANCED TECHNIQUES

Now that you have a basic understanding of TypePad, we're going to delve into some advanced techniques:

- Managing your blog's comments.
- Adding and removing categories.
- Adjusting your blog's design and About Page.
- Managing files uploaded to your account.
- Using feed technology for spam-free, virus-proof publishing.

Editing, Revising, Approving, and Removing Comments

Comments play a significant role in the success of blogs because they are an invitation to instant conversation. Comments can point you to more information on a post's topic. It's often through comments that you're directed to even more important information about a topic.

On occasion, you may want to edit, revise, or delete a comment left on your blog. I'll edit comments, correct misspellings, and remove inappropriate remarks or language where needed. (Like the time I had a neo-Nazi leave a link to a white supremacist Web site in a comment . . . ah, good times!) I'll also delete comments that are spammy or of little substance. Follow these steps to edit, revise, and remove comments on your blog:

1. Log in to TypePad and go to **Weblogs > (Your Blog)**. You'll be viewing the Manage tab and will

see that TypePad lists your most recent posts, as well as your blog's most recent comments.

2. To edit a comment, click the **comment excerpt** and a pop-up displays. Click **Edit**. An edit comment screen will load, similar to the edit post screen that allows you to revise, edit, or delete the comment.
3. To delete a comment, click the **checkbox** next to the comment and click the **Delete** button. To edit older comments go to **Weblogs > (Your Blog) > Manage > List Comments.**

Turning on Comment Approval

Some bloggers prefer to approve every comment before it appears on their blog, and comment approval allows you to do just that. To turn on comment approval:

1. Go to **Weblogs > (Your Blog) > Configure > Preferences.** The Preferences screen displays.
2. Scroll down to **Comment and TrackBack Preferences**.
3. Check the **checkbox** that says **Hold comments and TrackBacks for approval.**
4. **Save** Changes.

Now, each time you receive a comment or TrackBack, it is held for approval. TypePad will e-mail you to let you know that you need to log in to approve comments or TrackBacks. To approve comments:

1. Go to **Weblogs > (Your Blog) > Manage > List Comments**. Your most recent comments are listed. Comments that are awaiting approval are in bold.
2. To approve a comment, click the **comment excerpt**. A pop-up appears. Click **Publish**.
3. To approve multiple comments, check the **checkboxes** of the comments you want to approve and click **Approve**.

You can "un-approve" comments by clicking on the comment excerpt and selecting **Unpublish**.

Cultivating a Climate for Comments

Develop a habit of responding to every comment posted to your blog—put it in the blog itself, and send a separate thank you e-mail to the comment writer. If readers see you as responsive, they'll be more likely to continue to be a loyal readers and comment writers in the future.

Adding or Removing Categories

Every TypePad blog you create comes with a default set of categories to get you started. You may want to add custom categories, more specific to your blog's content, to better organize your entries. TypePad can include a list of your blog's categories in the side columns of your blog. This allows readers to quickly and easily navigate to predefined areas of interest.

If you are using your blog to write a book, use categories to catalog content for particular chapters. If you are running a seminar, use categories to organize course content for the day the content is covered. If you have multiple audiences for your blog, use categories to allow readers to self-select what they wish to read about. Then, if someone wants to read only your posts about marketing, they can click Marketing from your category listing. Here is how to add or remove categories:

1. To edit your categories, log in to TypePad and go to **Weblogs > (Your Blog) > Configure > Categories**. A list of TypePad's default categories displays.
2. To remove a default category, uncheck the checkbox next to it and click **Save Changes**.
3. To add a category, type the category name in one of the blank fields provided and click **Save Changes**. You'll only see space to add three categories at a time. Once these are added and saved, you'll see three more blank fields, allowing you to add more categories.
4. To remove a category you've added, click **Delete** to the right of it and click **Save Changes**. Removing a category will not remove posts in that category. They can still be accessed through the monthly archives.

Next time you add or edit a post, the revised category list will appear. You can add, remove, and edit categories at any time.

Change Your Blog's Design

TypePad comes with a selection of designs. A design includes the layout, fonts, and colors for your blog. The TypePad template engine allows you to manage all of these preferences with just a few clicks.

You may want your blog and main Web site to share common colors or fonts so the transition from one to the other isn't visually jarring. Or, you may want to make them very different if they have different content or different "voices."

A TypePad design is made up of five parts:

1. **Name and Description.** You can change the name of any of your designs for easy reference. Your blog's readers won't see that you named your design "Fall 2005 Design."

2. **Theme.** This is the fun one. A design's theme includes the font settings, borders, and other colorful elements of your blog.
3. **Layout.** This is the basic visual design of your blog's pages, including how many columns you have and where they appear. There are also advanced multimedia layouts for photographers and other visual artists.
4. **Content Selection.** You can choose among different content elements for what you wish to display in your overall blog's sidebar columns.
5. **Content Ordering**. After you choose what you're going to show, then you need to decide where you want it to appear on the page.

Changing a Design's Name and Description

To rename a design or change its description, go to the Design tab, click **Change Name and Description**. The Name and Description page displays. Revise or edit the name and description as you wish and click **Save Changes**.

Changing a Design's Theme

To change the color palette and fonts of a design:

1. From the **Design** tab, click **Change Theme**. The Theme page displays. TypePad comes with a set of predefined themes.
2. Select a Theme from the list provided and click **Save Changes**.
3. To see your newly themed blog, click **View Weblog**.

Changing a Design's Layout

To change the number of columns your blog has and their orientation:

1. Click **Change Layout**. The Layout page displays. You'll see a list of combinations of columns as well as some advanced multimedia layouts.
2. Select the Layout you want and click **Save Changes**.
3. To see the new Layout on your blog, click **View Weblog**.

Changing Content Selections

Once you've decided on the general look and feel of your blog, you then have to select the content that you want to show:

Click **Change Content Selections**. The Content Selections page loads with a list of all the various types of content you can show in your blog sidebar columns. Each item has a check box next to it. Choose the items you want to appear on your blog.

The first set of options includes information about **Your Blog Posts**:

Weblog Date Header: This groups posts by day, so if you have five posts on one day, they'll be under a header with that day's date.

Weblog Post Title: Select this option if you want the post's title to be displayed. This helps to improve your search engine visibility when key words are included in your post titles.

Weblog Post Footer: Check this box if you want a post footer to display at the bottom of each post, and to choose how you want it to be worded. If you want your Permalink to appear on your posts, you'll need to have a post footer.

The next set of options allows you to choose between various elements of **Sidebar Content:**

Amazon Wish List: This option is mostly useful for voyeur-cam teens who want presents from dirty old men, but can also be used to show your personal preferences in things like books, music, and movies.

Archive Links: A link to your blog's archives.

Categories: A list of your blog's categories. This enables readers to self-select the content they want to read.

Recent Posts: The most recent posts to your blog. This way, if a reader doesn't come to your blog through the home page, they can easily find and view your latest posts.

Recent Comments: A list of your blog's most recent comments with links to those discussions.

Side Calendar: A calendar of the current month. Days with posts are linked. I don't use the calendar because readers often confuse a calendar of posts with a calendar of events.

Your Photo Albums: This shows the most recent photo from each of your photo albums, linked to each individual photo album.

About Page Link: The links to your bio and profile.

Your Photograph: To upload your photo go to **Control Panel > Profile > Your Photograph**.

Online Status: Your status on any of the popular instant messaging programs such as Yahoo! Messenger, MSN Messenger, AIM, Skype, or ICQ. To specify your username for each of these services go to **Control Panel > Profile > Contact Information.**

E-mail Link: A link for readers to send you e-mail. This link is spam-proofed to prevent e-mail address harvesters from grabbing it.

Subscribe Link: Allows other bloggers to add you to their friends list, but only if they are also TypePad users.

Syndicate Link: The link to your blog's feeds. Feeds are discussed on page 140.

Podcast Link: The link to your blog's podcast. Podcasting is covered in greater detail on page 149.

Recent Updates: A list of the most recent TypePad

blogs updated. This may not be appropriate for a business blog since blogs with inappropriate or unrelated content may appear here. Your business blog might not benefit from having a link to an angst-ridden Poe-loving teenager.

Powered by TypePad Link: Adds a note to your blog saying how long you've been a TypePad user.

Your TypeLists: You can also choose from Your TypeLists. These are lists of your favorite people, places, and things. TypeLists are covered in greater detail later in the section "Creating and Adding TypeLists," on page 129.

Your Feeds: You can show the most recent headlines from your favorite news sites or blogs by adding their feeds here. If you have more than one blog, this is a way to cross-link them. Feeds are discussed in detail on page 140.

Your Photo Albums: You can specify which of your photo albums you want to show in your sidebar columns.

Once you've made your content selections, you can **Preview** to see how the blog will look or click **Save Changes**.

Changing Content Ordering

Now that you have selected what content you are going to display on your blog, you can specify the order in which you want it to appear:

1. Go to **Weblogs > (Your Blog) > Design > Change Ordering**. The Ordering screen displays.
2. You can now click, drag, and drop the sidebar elements in the order and in the columns in which you wish them to appear.
3. Click **Preview** or **Save Changes**.

After changing any of these options, you can click **View Weblog** to see your updates.

SUCCESS STORY

BLOGGING WITH HEART

Pat Gundry, a writer with six published books, found blogs to be the best way to promote her books, stay in contact with readers, and manage new projects. She's used the blog format to focus on a controversial topic close to her heart: making abortion unnecessary.

For years, I've had a big project waiting in the wings, nudging me from time to time to get started on it. But there were always so many hoops to jump through, and the cost was prohibitive. The project would involve the creation of a book, interviewing many people, and some broad means of public news dispersal—the most difficult and expensive part of the project. Then blogs appeared, and almost immediately I realized I could now take my big project live. A blog was the key to making it happen. Gone was my need for a big PR campaign, costing me big bucks. I created my blog and began my work on the "What If?" campaign. I'm interviewing subjects and writing the book now, looking forward to adding my efforts in solving, in an ecological way, a large and troubling social problem.

You can read one of Pat's many blogs at www.makeabortionunnecessary.com.

Using Custom Themes

The standard design themes simplify design decisions, but what if you want to have a blog with a different color palette or font style? TypePad allows you to tweak the theme of a design to suit your whims and preferences.

Go to **Weblogs** > **(Your Blog)** > **Design** and click **Change Theme**. The Theme page displays.

Before, we had used Predefined Themes and selected from a number of predefined designs. This time, instead, select **Custom Theme**. A whole new set of options display:

General Page Settings: You can select the colors and borders of the overall page and columns, as well as column width, and the colors of your blog's links.

Page Banner: This option allows you to adjust the

color, font, and style of your blog's title and tag line on the page. You can also upload an image if you want to use a graphic banner. This is a great way to align your blog's look and feel with your existing Web site or marketing materials.

Weblog Posts: In this area you can specify the style of your blog's posts, including font, style, size, color, and line spacing. Likewise, you can style the post title, the post body text, and footer.

Sidebar Items: The style components of the items in the blog's sidebar columns can be adjusted, including font, style, and color. The style of any links appearing in the columns can be similarly adjusted, as well.

Click **Edit This Element** to edit any of these sets of options and a pop-up window will open, allowing you to customize that set of options. You'll be able to **Preview** or **Save Changes** to update your blog's design.

Tweaking Your Profile and About Page

Your About Page is important because it allows others to get to know you without reading through your archives. By showcasing your professional skills, as well as your personal side, you'll be more memorable to readers.

Here's how you can change and customize the information that appears on your About Page:

1. Go to **Control Panel** > **Profile**. The Profile screen displays.
2. Fill in whatever information you want to share with others using the following fields:

   Upload Your Photo: Under *Your Photograph*, click **Browse**. Locate your picture on your hard drive and select it.

   Password Settings: Here is where you change the password for your TypePad account.

Contact Information: If you use instant messaging services, you can add your aliases here.

Miscellaneous Information: If you are an Amazon Affiliate, you can enter that information here (to join, go to associates.amazon.com). If you have a wish list on Amazon, you can indicate that here as well.

To find your Amazon Wishlist ID number, go to http://help.typepad.com/panel/profile.html#wishlist

Your Interests, Your Extended Biography, Your One-Line Biography: Use these fields to add a little personal flavor to your About Page. You can list your formal business credentials and experience, but it might be good to include hobbies and quirks as well for that added touch.

3. Click **Update Author Profile**.

Customize Your About Page

Now that you've filled out the fields in your Profile, you can decide what you want to make public on your About Page. For example, you may not want to show your nickname, but may want to show your main Web site's URL.

1. Go to **Control Panel > Profile > About Page.**
2. Check the boxes of the information you want to show on your About Page.
3. You can also **change the style** of your About Page.
4. Click **Save and Publish**. Your About Page is updated instantly.

Note: You can only have one About Page for your TypePad account. If you have multiple blogs, they'll all have the same About Page. So, if you have your party-skank tell-all personal blog and your more formal business blog, you'll need to design an About Page to serve both.

Advanced Posting Options

TypePad users who subscribe to premium services can add additional posting options, which add more flexibility and detail to their blog's posts. To see these options:

1. Go to **Weblogs > (Your Blog) > Post**. At the bottom of the post editing screen is a link that says **Customize the display of this page**. Click that link and a pop-up window opens showing additional options:

   Post Editor: You can choose between the fancy "Rich Text" WYSIWYG editor or a traditional, plain-text editor.

   Post Screen Configuration: The Basic option is what you start with. TypePad Basic users are limited to this configuration. **Custom fields include:**

Post Continuation: This feature allows you to break a post up into two sections so only the post's introduction shows up on your home page. After the introduction is a "Continue Reading" link that leads to the archive page for just that particular entry, where the entire post will display.

Excerpt: Excerpts of your posts can be used in your syndicated feeds to tease subscribers to read the entire entry on your blog.

Key words: Key words, like Excerpts, make your blog entries more palatable to search engines.

Text Formatting: By default, TypePad converts all line breaks (like when you press Enter or Return between lines) into new paragraphs. This can be annoying when you're trying to do more advanced HTML. This feature allows you to customize your formatting, sidestepping this annoyance.

TrackBacks: TrackBack is a feature that enables easier cross-linking between blogs and online conversations.

You can also choose where you want the **Preview**, **Save**, and **Delete** buttons to appear—at the top or bottom of the page.

2. Check the options you want, and Click **Save**. The pop-up window closes and the updated post screen displays.

SUCCESS STORY

WRITE YOUR BOOK

Andrea J. Lee is an award-winning author and consultant whose focus is helping small business owners make meaning and money sustainably through multiple streams of active and passive income:

I wrote my book using a private blog. Without it, I doubt I could have written it within my self-imposed ninety-day deadline. I needed a tool to help me get cracking without going insane. Specifically, the blog structure allowed me to pour the content of the book out of my brain and into a structure. Each post was a chapter and I was able to re-arrange chapters cleanly and easily, so that I could see, at a glance, how the "bones" of my book were looking. By the time I reached thirty posts, I knew I was done. I could solicit comments from editors, and enlist collaborators to contribute success stories, all through the blog environment. It reduced the stress of "managing" the birth of a book, much like a mind-mapping tool. Want to write a book fast, with minimal fuss, and start selling it sooner than you thought possible? Get a blog . . . it's the only way to do it. I'm working on two more book projects right now where I'm doing the same thing!

You can read Andrea's blog at www.andreajlee.com.

Posting by E-mail, Cell Phone, or Camera Phone

Mobile blogging (or "moblogging" if you want to sound like a complete geek) allows you to make a post even when you're not at a computer. The easiest way to do this is through e-mail. You send an e-mail to TypePad and it automatically formats it into a blog post. If you send a photo or image, it can even post it to one of your photo albums.

1. Log in to TypePad and go to **Control Panel > Profile > Mobile Settings**. The Mobile Settings screen displays.
2. **Select your default posting settings.** You can designate a particular blog and photo album to which you can send posts and photos. You can also specify that you want to receive a confirmation e-mail of receipt. If you are billed for

every incoming message on your cell phone, you may want to forego this option.

3. **Indicate the e-mail address you'll be using.** Tell TypePad what e-mail address your posts will be coming from. For example, if you are posting from work, enter your work e-mail address.
4. **Select how you would like the system to verify your messages.** TypePad will want to ensure that the e-mail coming from you is really from you, so there are three verification options:

   Secret E-mail Address. TypePad will generate a unique e-mail address to which you can send e-mail. The idea is that it's too long of an address for someone to just guess. This is the easiest option.

   Confirmation Messages: TypePad will send a confirmation message to ensure that it really was you sending the e-mail.

   PGP Signatures: PGP stands for (I kid you not) "Pretty Good Privacy" and is the industry-wide standard for encryption. Unless you know what a PGP signature is, you probably don't need to worry about this one.
5. Click **Save**. Your changes are saved.

6. **Test the setup by sending a text e-mail to the e-mail address.** Within a few minutes, the e-mail should be posted to your blog. Notice that the subject line of your e-mail becomes the title of your post, while the body of the e-mail becomes the post.

Managing Your Files

Aside from uploading images and files when writing a post, you can also manage your TypePad account's files through the **File Manager** feature. So, if you upload three images while composing a post and then only use one of them, you can delete the others.

To access the File Manager, log in to TypePad and go to **Control Panel > Files**. The **File Manager** displays. You'll see a folder for each blog or photo album on your TypePad account. Click the name of a folder to view that folder's contents.

To delete a file or directory, check its check box and click **Delete**.

To upload a file to the current directory, use the **Upload a New File** form to locate the file on your computer and then upload it.

To create a new folder in the current directory, use the **Create a New Folder** form.

You'll notice that some directories or files cannot be deleted. They are essential to your blog's structure and cannot be removed.

* * *

In part four, we looked at some advanced techniques to help you take full advantage of TypePad's features. You learned how to manage files, edit comments, send posts by e-mail, and also to adjust your blog's design to suit your tastes or to match existing marketing materials. Now that your blog is out there in the blogosphere, we'll take a look at some of the ways you can use it to promote your business.

PART FIVE

PROMOTING YOUR BLOG

In the final part of this book, we'll discuss how to promote your blog and increase your connectivity in the blogosphere. We will look at:

- How to use TypeLists to catalog and link to your favorite sites and bloggers.
- How to monitor your blog traffic to see where visitors are coming from and what they're looking for.
- How to use the pinging option.
- How to integrate your blog into your existing marketing efforts.

Creating and Adding TypeLists

A *TypeList*, as the name implies, is a list of things of a certain type: a reading list, a list of your favorite Web sites, a list of your favorite musicians, or a list of whatever you want (documents, events, rules, quotations, etc.). TypePad allows you to organize lists of things quickly and easily without having to know any HTML. Your TypeLists can be used for anything you want to appear in the sidebars of your blog. They're effective because they give the reader a quick snapshot of you, your business, your interests, and your products and services. TypeLists are especially helpful in managing and organizing a blogroll, which is a list of that blogger's favorite blogs and Web sites.

Create a TypeList

1. Click the **TypeLists** tab, one of the four main tabs at the top of the screen. TypePad then lists your current TypeLists.
2. Under **Create a New TypeList**, select the type of TypeList you want to create. You have several options:

   People: TypePad will prompt you for the URL of the people you want on your list. Most TypePad users use this option to manage their blogroll.

   Link: As with the People TypeList, TypePad will prompt you for the URL of the site you want on your list, then it will try to grab the name of the Web site from the Internet.

   Reading: TypePad will prompt you for the ISBN numbers of the books you wish to list and then query Amazon.com to grab the title information. Alternately, you can enter a title and TypePad will attempt to guess which book you are trying to add.

   Music: With this option, TypePad makes a list that you'll then have to update manually.

Notes: The Notes feature of TypeList allows you to add free-form text or advanced HTML to your side column content. Use this when adding newsletter subscription forms or other swatches of HTML that don't fit into a list format.

When in doubt, start with a Link list.

3. After you create the TypeList, TypePad prompts you to add it to any of your existing blogs. A quick pop-up will appear allowing you to select which blogs you want the TypeList to appear on, including your About page.
4. Click the **Add a new item** link to add a new item to your TypeList. If your list is a Reading list, you can add books by entering their ISBN code or by entering search terms, and TypePad will then guess the book's title by searching Amazon.com's database. If you've specified your Amazon Associates ID in your Profile, then TypePad will make the link an affiliate link.

TypeLists can be used to add lists of documents, events, rules, quotes . . . anything you want to put in the side columns of your blog.

Monitoring Your Blog's Traffic

Just as with a conventional Web site, there are several ways you can measure who is reading your blog, what search engines are finding you, and where you're being talked about. This data can be used to gauge your blog's success, to adjust your content, and to network with other bloggers.

Built-in Stats

TypePad includes basic traffic reports that allow you to see where your blog's traffic is coming from. To access the built-in stats, go to **Control Panel > Stats**. Use the drop-down menu at the top to select what blog or photo album's traffic you would like to view.

The report includes the number of hits by week, day, and hour, as well as a detailed list of what pages are being hit and where the traffic is coming from. In my example from the introduction, I was able to tell that readers were finding my blog by searching for "John Kerry Herman Munster" in various search engines.

Notice that if traffic is coming from a search engine like Google or Yahoo!, you can just click the link to see what search terms are sending people to your blog. This valuable information will help you align your blog to certain key words or search terms. If a hit has no referring link, it probably came from a bookmark, an e-mail, or someone simply typing in your blog's address.

Site Meter

Site Meter (www.sitemeter.com) is a free service where through a small swatch of code you place on your blog, this service can cook up reports on your daily traffic and where it's coming from. Stats reported include referring links, geographic location of readers, language, and busy times of day. This useful information supplements the existing statistics available in TypePad. Similar services are HitBox (www.hitbox.com) and StatCounter (www.statcounter.com).

Technorati, Feedster, and BlogPulse

Technorati (www.technorati.com), Feedster (www.feedster.com), and BlogPulse (www.blogpulse.com) are blogging ecosystems that monitor millions of blogs and what they're all talking about. You can use these sites to find out when another blog or site is linking to you. Submit your blogs to these sites and within a day or so you'll be able to search the services to see where your blog is nestled in the blogosphere.

Password Protecting Your Blog

TypePad allows you to password protect individual blogs, photo albums, or your entire TypePad site. This is a great tool for businesses that want to use blogs internally or publish a "clients-only" blog.

There is one drawback: You can only have one user name/password combination for each blog. Because the passwords are not super-duper secure, this is probably not the best approach if you're using a blog to manage development of the next generation of nuclear weaponry, but they're probably good enough for most uses. To set up password protection:

1. Go to **Control Panel > Site Access > Password Protection.** The password protection screen displays.
2. You have three options:

- **No password protection**.
- **Protect the entire site** (all blogs, photo albums, feeds) with one user name and password.
- **Protect certain areas of your site.** When you select this option, a list of your blogs and photo albums will drop down with a check box to password protect certain areas of the site, along with fields to enter user names and passwords.

3. Click **Save Changes**. The password protection is immediately activated.

SUCCESS STORY

EFFORTLESS MARKETING

Julia Stewart is a certified mentor coach focusing on mentoring budding coaches on building their coaching practices:

Blogging brings me clients effortlessly. It's not unusual for me to get e-mails from people saying, "I love your blog. It speaks to me. I want to work with you!" I also get calls and e-mails from colleagues who want to collaborate with me on intriguing projects. So it's increasing the number of opportunities that come to me. My blog has also been quoted by other publications, which brings me more exposure than I could pay for. All in all, I'd say there's an added degree of electricity around my brand, since I started blogging. And I'm really just writing what I want about stuff that has to do with coaching. It's as effortless as marketing can get. Kind of like non-marketing marketing!

You can read Julia's blog at juliastewart.blogspot.com.

Turning on Third-Party Notification

Weblogs.com (www.weblogs.com) and blo.gs (www.blo.gs) are two Web sites that serve as clearinghouses for updated blogs. Each time you add a post to your blog, TypePad can notify these two services that you've added new content. This is called *pinging*. It is like ringing a dinner bell and shouting "Hey! New posts! Come and get 'em!" Other services and search engines use this data to let their users know when certain blogs have been updated.

You'll see this data in action when you view a blog's blogroll and see a little note next to certain blogs, meaning they've been updated in the last twenty-four hours.

To turn on notification:

1. Go to **Weblogs > (Your Blog) > Configure > Publicity & Feeds**. The Publicity & Feeds screen displays.
2. Under **Would you like to notify third-party services when you update your weblog?** check the two check boxes.
3. Click **Save Changes**. That's it!

For other pinging and update services, check out Ping-O-Matic (www.pingomatic.com), Pingoat (www.pingoat.com), and King Ping (www.kping.com). TypePad doesn't support these services directly, but you can still use them to ping the appropriate servers.

What Is a Feed and How Does It Work?

A *feed* is a small file containing a Web site's latest updates and headlines, which is updated every time the site itself is updated. Feeds are called many things: RSS feeds, XML feeds, Atom feeds, news feeds, or Web feeds. To keep things simple, I'll just call them feeds.

Feeds allow your blog's updates to be received by subscribers through a spam-free, virus-proof delivery channel that bypasses their e-mail inbox.

Just as a word processing document can be in Microsoft Word or OpenOffice.org format, a feed can also be published in multiple formats. Think of them as flavors. Common formats are *RSS* and *Atom*.

RSS is the granddaddy of feed formats and stands for "Really Simple Syndication." Geek-freaks debate what it really stands for and many arguments have resulted. Nearly every blog platform can create feeds in RSS format.

Atom is a newer format that was built from the ground up by blog tool programmers frustrated with the RSS format runaround. Atom has more options than RSS and is gaining wider adoption among blog platforms.

Nearly all feed formats are in an overarching format called XML. That's why sometimes a feed is referred to as an XML feed.

A *news aggregator* is the software that goes and grabs the feeds of your favorite sites and tosses them into your own customized newspaper. News aggregators can be Web-based, like Bloglines (www.bloglines.com), or actually run on your computer, like Pluck (www.pluck.com) or Newsgator (www.newsgator.com). Using a news aggregator is like having your own private assistant chained to a research desk.

How Do Feeds Work?

Here is an overview of how feeds work:

1. A blogger posts new content to his blog.
2. As the blog is updated, the blog platform

updates the site's feed with the blog's latest headlines and posts.

3. The subscriber's news aggregator, which is subscribed to that particular blog, notices that there are updates to the feed.
4. The news aggregator downloads the latest headlines and posts for the blog and makes them available to read.
5. The subscriber can then view the blog's most recent posts along with the rest of his or her favorite news and updates in a customized newspaper.

To view an animated demo of this process go to http://www.goblogwild.com/goodies/.

Turn on Your Blog's Feeds

To turn on syndication, which allows other people to read your site through their news aggregators:

1. Go to **Weblogs > (Your Blog) > Configure > Publicity & Feeds.** The **Publicity & Feeds** screen displays.
2. Under **Feeds**, select an option: I recommend **Yes, offer a feed with a short excerpt from my**

posts. This will publish a short teaser or excerpt of each of your posts and will bring the reader to your site to read the whole post if they are interested.
3. Click **Save Changes**.

After making this selection, you should see a link that says **Subscribe to this blog's feeds** in your blog's sidebar columns. That link is going to be the feed that subscribers add to their aggregators. If you don't see the feed link, go to **Weblogs > (Your Blog) > Design > Change Content Selections** and be sure **Syndicate Link** is checked.

Start Using a News Aggregator

The fastest way to understand how feeds work from the subscriber's perspective is to start using a news aggregator. Here is a walk-through for using Bloglines, one of the more popular Web-based news aggregators.

First, sign up for Bloglines at www.bloglines.com. You'll fill out a registration form to create your Bloglines account. Once you've confirmed your registration, you can log in to Bloglines. Like TypePad, Bloglines has a tabbed interface to facilitate navigation.

Click the **My Feeds** tab. The My Feeds screen displays. The My Feeds window is split into two "panes":

a) The left pane shows the feeds you are subscribed to.
b) The right pane shows the latest updates for that feed.

By default you have been subscribed to the feed for Bloglines News. Click **Bloglines News**. In the right pane the latest updates from the Bloglines News feed displays. Each entry includes a linked headline and an excerpt of the news story or article. Click the headline if you want to see the news article on the actual Bloglines News blog.

You'll also see an **Unsubscribe** link that allows you to remove a feed subscription.

Subscribe to a Feed

Let's walk through the process of subscribing to a feed in Bloglines:

1. Click the **Directory** tab. The Directory displays.
2. Click **Most Popular Feeds**. The list of the most popular feeds displays.

3. Find a feed you like. Click **Subscribe**. The details and options for this feed display.
4. Don't worry about all the options available for right now. Click the **Subscribe** button. The **My Feeds** page displays with the new feed added to the list on the left pane.
5. Click the linked name of the feed, and the latest updates display in the right pane.

You can find other feeds in the Bloglines directory using the search function.

Find Feeds

There are many places to find feeds for your favorite blogs and Web sites.

On the Web site itself: Most sites have buttons or graphics that link to the site's feeds. They usually look like an orange-and-white button with the words RSS or XML. Looking for the words *RSS, Atom, XML*, or *Subscribe* is usually a pretty good tip-off. To get the link of a feed, click the actual button, link, or graphic. You can view the link to the feed in your browser's address bar.

In a feed directory: There are many feed directories. Some of the more popular ones are Feedster (www.feedster.com), Syndic8 (www.syndic8.com), and News Is Free (www.newsisfree.com). Once you navigate to a certain news source, you'll see the familiar orange-and-white button or link. Click the link to see the feed.

Google it: Some of the bigger news sites aren't really that user-friendly when it comes to finding out about their feeds. In Google, search for the name of your news source and the letters *RSS*.

What to do with the feed link: When you click a feed button or link, it might look like techie gobbledy-geek. But all you need is the link located in your browser's address bar. Just copy and paste that link into your news aggregator.

The Big Deal about Feeds

Now that you've started using feeds, you've probably noticed a few distinctions between feeds and e-mail newsletters.

Feeds are spam-proof. Evil spammers have made everybody less likely than ever to subscribe to e-mail newsletters.

Feeds are virus-proof. E-mail-based viruses have eroded the trust in e-mail delivery. Feeds protect your subscribers from e-mail-based viruses and worms.

Feeds put the power in the recipient's hands. Give out your e-mail address once to the wrong person, and it can be abused indefinitely. A feed is like a TV channel: if it becomes annoying, you can just turn it off for good.

Feeds increase your content's value. Your messages are on par with the recipient's favorite and most important news. You are important enough to be a part of their customized newspaper. Instead of being an interruption in their daily reading, you're front and center.

Feeds help your search engine ranking. Search engines like Yahoo! and Google use feeds as a faster way to index a site's content and updates. Putting the feeds of other sites on your own blog can also keep your content fresh. You can do this with TypePad. Go to **Weblogs > (Your Blog) > Design > Change Content Selections > Feeds** and click **Add a new feed**.

Feeds can not only make your site more attractive to search engines, but can also be used through a news aggregator to create a customized view of the information you want to know about. In addition, they can help you research, publish, and market your business online.

For a more in-depth exploration of feeds and related technologies, check out RSS Essentials (www.rssessentials.com).

What Is Podcasting?

With feeds, news aggregators are usually just grabbing text and pictures. What if you could expand this to include audio? Enter podcasting.

Podcasting works just like a feed, downloading updates automatically just like a news aggregator. But with a podcast you are downloading audio instead of text. The *pod* part comes in because you can then synchronize these audio files to your iPod or other MP3 player, though you don't have to have an iPod or MP3 player to enjoy podcasts.

Just as blogs have revolutionized online publishing by giving anyone, anywhere the tools to be read, podcasting gives anyone, anywhere the tools to be heard and seen (podcasting also supports video).

To get started listening to podcasts, you'll need to download a podcast receiver like Juice (www.juice

receiver.com) or the latest version of iTunes (www.itunes.com).

If you are interested in learning how to create your own podcasts and start your own Internet radio show, visit the Podcasting Bootcamp (www.podcastingbootcamp.com).

Promoting Your Blog

Once you're up and blogging, how do people find your blog? Just starting a blog isn't going to bring you success. You must cultivate your blog to grow your audience market.

Here are some simple ways to get traffic to your blog:

Tell customers, clients, and colleagues about your blog. Include a reference to your blog in your e-mail signature, newsletter, discussion forum profile, business cards, and any other marketing materials. Don't forget the value of traditional press releases that you can submit to an online submission service like PR Web (www.prweb.com). The more you make your blog a central part of your marketing strategy, the better.

Get indexed by the blog search engines as well as conventional search engines like Yahoo! and Google.

Robin Good of Master New Media (www.masternewmedia.org) has a constantly updated list of places to list your blog. His list is focused on listing your blog's feeds, however most of the sites apply to the blogs themselves.

You can find Robin's list at http://www.masternewmedia.org/rss/top55/.

Always be linking. Link to other resources and blogs that you like to read and that are related to your topic. Become a regular on someone else's blog—they might become a regular on yours.

Remember the power of key words and phrases. Which search terms might point people to your products and services? Use tools like Wordtracker (www.wordtracker.com) to explore which search terms potential customers might use to find you. Write a post with one of those terms as a title (examples: "Tax Accountants in Chicago" or "Single-Mom Resources"). You can then post reviews of books and products related to your blog's focus.

The Best Way to Get Blog Traffic

The best way to generate traffic to your blog is by commenting on other blogs. Blogging is about conversation and you'll need to get into the habit of reading other blogs in your field or industry and commenting on those blogs, while always linking back to your own blog. Agree, rebut, expound, or expand on the substance of what the blogger is saying. This process is not about commenting on every blog you see (that would be spamming), but about finding other bloggers in your field with whom you can collaborate and converse. When commenting on other blogs, take advantage of the URL or Web site field in the comments form by typing in your blog's URL there. Then, if a reader finds your comment interesting, he or she might just click over to read more about you on your blog.

No one is going to be a part of your conversation until you are a part of theirs. Give your blog time to build. A blog is not some fly-by-night automated scheme; it's a vehicle for sharing your passion for life and business with others. It's a means of building a reputation that leads your readers to become your customers.

What's Next?

Whew!

With all there is to learn and know about blogging, it's easy to feel overwhelmed. Here are some parting tips to keep everything in perspective.

Blog on a regular basis. Make blogging an integral part of the way you run your business. Start blogging Web sites instead of bookmarking them and point customers to resources on your blog. Make your blog a conduit for others to contact you.

Flaunt your quirks. Your eccentricities are what will set your blog apart from all others. Show some personality, reveal your passions, and demonstrate your expertise. Blog about that book you read on vacation, while also sharing pictures of you and your daughter building a sand castle. Stop hiding behind a monolithic brand and show the world who you really are.

Get started. There's no right time to start a blog. Just go to www.typepad.com and start experimenting. There are few "right answers" when it comes to blogging. Start finding out what blogging can do for you. Start where you are.

Have fun. If blogging isn't fun, you won't do it. Maintain a sense of possibility as you explore blogging and learn this new technology. Keep it light. Share your interests and your challenges. You never know what connections you'll make online.

You have everything you need to get started. You can take this technology in any direction you desire. By combining your creativity with blogging technology, you can communicate with the world. But first, one more success story . . .

Instant, Global Self-Expression

The final success story in this book is my own. I've always been captivated by technology that allows us to self-publish, whether it was a fourth-grade puppet theater, play production in Chicago, or blogging, with its ability to reach a worldwide audience. It's been exciting to see blogging go from the Internet underground to the evening news and beyond.

I started teaching blogging to entrepreneurs because it was a way for them to share the stories of their businesses, and their lives, on a global scale. Blogging delivers on the promise of the Internet—anyone, anywhere can say anything, and anyone, anywhere can read it. You never know who's going to read your blog.

I was checking my e-mail a year ago, when I got a message from an editor at Portfolio publishing. She had seen my blog, liked my writing style, and asked if I would be interested in publishing a book. Who would say no to that question? I didn't have an agent. I didn't even have a book proposal. A couple of days later, she came back with a formal offer for publishing a book worldwide.

I was not the most widely read or influential business blogger out there, but my individual voice set me apart, and my archives showcased my knowledge.

So yes, as strange as it might sound, blogging can change your life. It teaches you about yourself, and creates community and connections you never thought possible. Blogging is a way to chronicle not just the story of your business, but also the story of you. This is what customers everywhere want to hear. A simple, honest human voice.

You can read my professional blog at www.andywibbels.com and my personal blog at www.andymatic.com.

Share Your Success

I've featured just a few success stories in this book.

I would like to hear from you! How has this book helped you get started with blogging? How have you integrated blogs into your business and personal life—and what were the results?

I want to hear your success story. Tell me about it at www.goblogwild.com/successstory/.

The Goody Bag

As in all good seminars, it's great to give participants a bag of goodies to take home with them. I wasn't able to shoehorn everything possible about blogging into this book, and some of the techie stuff might go out-of-date as blogging matures. So, for a complete resource of updates, tutorials, and other goodies go to http://www.goblogwild.com/goodies/.

Glossary

Archives—The past posts of a blog, organized by month, week, or category.

Atom—A popular feed format.

Blog—Short for "Web log." An easily, instantly, and frequently updated Web site, focused around a topic, industry, or personality.

Blogosphere—The collective hive of all blogs on the internet.

Blog platform—The software that manages a blog's content and publishing. Examples: TypePad and Blogger.

Citizen journalism—The practice of bloggers enriching stories in the mainstream media with their own coverage and views.

Comments—Responses to a blog post submitted by a blog's readers or written in response to readers by the blog's author.

Ecosystems—Services that monitor millions of blogs and are able to report on the most popular links and search terms used in the blogosphere.

Feed—A small file containing a blog's latest updates. Feeds are read by a news aggregator.

Moblogging—Short for "mobile blogging" and includes practices such as blogging by cell phone or camera phone.

News aggregator—A Web site or application that collates feeds into a customized newspaper. Also called a news reader, feed reader, or RSS aggregator. Example: Bloglines.

Permalink—A permanent and direct link to a particular blog post.

Ping—A notification sent to various online services that a blog has been updated with new content.

Podcast—A distribution method for Internet audio and video through direct download.

Post—An entry in a blog.

RSS—A popular feed format. RSS stands for Really Simple Syndication.

TrackBack—A form of remote commenting that automatically links a blog post to a post on another blog.

Vlog—A video blog with video posts instead of text.

XML—The overall family of formats that feeds are a part of .XML stands for "Extensible Markup Language."

Index

About the Author

Andy Wibbels is a blogger, speaker, consultant, and generally creative malcontent. His award-winning professional blog is at www.andywibbels.com and his ranty personal blog, Andymatic, is at www.andymatic.com. He was a contributing author to *Success Secrets of the Online Marketing Superstars* as well as the creator of the Easy Bake Weblogs, RSS Essentials, WordPress Essentials, and Podcasting Bootcamp seminars. He currently lives in Chicago.

You can e-mail him at andy@goblogwild.com.